Noshin Samiha Khan Trisha

Hypertension and nutrition

GRIN Verlag

Bibliografische Information der Deutschen Nationalbibliothek:

Die Deutsche Bibliothek verzeichnet diese Publikation in der Deutschen National-
bibliografie; detaillierte bibliografische Daten sind im Internet über http://dnb.d-
nb.de/ abrufbar.

Imprint:

Copyright © 2013 GRIN Verlag GmbH
Druck und Bindung: Books on Demand GmbH, Norderstedt Germany
ISBN: 978-3-656-64379-1

This book at GRIN:

http://www.grin.com/en/e-book/272338/hypertension-and-nutrition

GRIN - Your knowledge has value

Der GRIN Verlag publiziert seit 1998 wissenschaftliche Arbeiten von Studenten, Hochschullehrern und anderen Akademikern als eBook und gedrucktes Buch. Die Verlagswebsite www.grin.com ist die ideale Plattform zur Veröffentlichung von Hausarbeiten, Abschlussarbeiten, wissenschaftlichen Aufsätzen, Dissertationen und Fachbüchern.

Visit us on the internet:

http://www.grin.com/

http://www.facebook.com/grincom

http://www.twitter.com/grin_com

Abstract

Introduction: Hypertension is the most common risk factor for stroke, heart disease and also many other disease. African Americans are mostly vulnerable for hypertension. They have increased likelihood for developing hypertension. But good news is there are many treatment or remediation methods available for prevent and/or control hypertension. Dash Diet is the one of the methods to control hypertension.

Method: Hinari, Medline through PubMed and Google scholar was used for literature searching. Key words like hypertension, nutrition and hypertension, DASH diet, DASH diet and hypertension was used.

Results: There are no actual causes for increased risk of hypertension among African American found. Research shown that DASH diet has significant impact on reducing hypertension

Introduction

Hypertension or high blood pressure is also named as arterial hypertension. It is a chronic medical condition. Hypertension is now a day is very common in almost every region of the world. Hypertension is the result of elevated pressure of blood in heart. In case of hypertension heart works harder than the normal range to circulate blood through the blood vessels. Normal blood pressure is 100-40mmHg, here the 100is the systolic and 40 is diastolic. Normal blood pressure is 100-140mmHg systolic and 60-90mmHg the diastolic pressure (Hall, 2010). Hypertension is the most common risk factor for stroke, myocardial infraction, and heart failure, aneurysm of blood vessels, peripheral arterial disease and chronic kidney diseases.

Hypertension among African American is very common and hypertension develops at the younger age among the African American rather than the other groups in the USA (Ford et al., 2002). There are no actual data for why hypertension is common in African American but some preliminary studies showed that race may be the important factor for elevated blood pressure in African American. Another potential cause for elevated blood pressure among African American is anger inhibition (Steffen et al., 2003). As there are several articles about the cause of prominent hypertension in African American but there is no such data for actual cause. All the causes are derived from hypothesis. Cooper and Rotimi, 1997 stated that the underlying assumption that there is some genetically determined physiologic difference (Cooper and Rotimi, 1997). It is clear from various data that without some rare exceptions hypertension determines by several genetic factors. This genetic factor accounts more than 50% of the hypertension. So may be the genetical and physiological factors are responsible for increased hypertension among

African- American. From an current study conducted in 2006 stated a hypothesis that creatine kinase, which has been observed to be present in greater amounts in the skeletal muscle of African Americans than Caucasians, could be a genetic factor that predisposes African Americans to hypertension (Brewster et al., 2006). The good news is that treatment can prevent and control hypertension. There are several treatment method available includes losing weight, increasing physical activity. Following a healthy eating plan, which emphasis on fruits, vegetables, and lowfat dairy foods, choosing and preparing food with less salt and sodium can help to control hypertension (Campbell and Chockalingam, 1995). The Dietary Approaches to Stop Hypertension (DASH) diet was born from an initiative of the National Heart, Lung, and Blood Institute (NHLBI) to examine dietary factors that affect blood pressure. It had long been known that populations who ate diets based on vegetable products had lower blood pressure levels than usually found in western countries, and lower incidences of hypertension and stroke (Sacks et al., 1999).

This paper is aiming to review the importance of delivering nutritional education to the people with hypertension. This paper will focus on African American people context and will also explore the effectiveness of "DASH diet".

Method

In this review four database used for searching related literature, Google scholar, Hinari, Medline through PubMed. Searches were conducted by using key words- Hypertension, Nutrition and hypertension, DASH diet and hypertension. Google scholar and Medline through PubMed was the primary database. Hinari used to find out the articles according to their journal if full text was not available on Google scholar and Medline search.

Result

DASH diet

The DASH diet (Dietary Approaches to prevent Hypertension) may be a dietary pattern promoted by the U.S.-based National Heart, Lung, and Blood Institute (part of the National Institutes of Health, bureau of the united states Department of Health and Human Services) to forestall and management cardiovascular disease. The DASH diet is made in fruits, vegetables, whole grains, and low-fat farm foods; includes meat, fish, poultry, balmy and beans; and is restricted in sugar-sweetened foods and beverages, red meat, and additional fats. Additionally to its result on pressure, it's designed to be a well-balanced approach to feeding for the overall public (National Heart, 2012, Appel et al., 2006). It's currently counseled by the US Department of Agriculture (USDA) as a perfect feeding set up for all Americans(Sacks et al., 1999).

The DASH diet relies on authority studies that examined 3 dietary plans and their results. None of the plans was eater, however the DASH set up incorporated a lot of fruits and vegetables, low fat or skim farm, beans, and balmy than the others studied. The diet reduced systolic pressure by half-dozen mm hg and diastolic pressure by three mm hg in patients with high normal pressure, currently known as "pre-hypertension." Those with cardiovascular disease born by eleven and half-dozen, severally. These changes in pressure occurred with no changes in weight. The DASH dietary pattern is adjusted supported daily caloric intake starting from 1600 to 3100 dietary calories (National Heart, 2012, Appel et al., 2006, Sacks et al., 1999).

Daily Nutrient Goals Used in the DASH Studies (for a 2,000-Calorie Eating Plan) (National Heart, 2012)

Total fat	27% of calories
Saturated fat	6% of calories
Protein	18% of calories
Carbohydrate	55% of calories
Cholesterol	150 mg
Sodium	2,300 mg*
Potassium	4,700 mg
Calcium	1,250 mg
Magnesium	500 mg
Fiber	30 g

* 1,500 mg of sodium was a lower goal tested and found to be even better for lowering blood pressure. It worked very well for people who already had high blood pressure, African Americans, and middle-aged and older adults (National Heart, 2012).

g = grams; mg = milligrams

Fruit and Dash Diet

The "Fruits and Vegetables'' diet arm proved that fruits and vegetables, as well as wacky, lower blood pressure and are liable for a minimum of 1/2 the whole impact of the DASH diet. Fruits, vegetables and wacky are high in potassium, magnesium, fiber, and plenty of different nutrients. Of that potassium is best established for lowering vital sign, significantly in persons with low intake, in hypertensive persons, and in African Americans. The DASH diet multiplied potassium intake from an occasional daily quantity of roughly 1,700 mg to a high level of four, 100 mg (Sacks et al., 1999). The magnitude of the vital sign reduction on the fruits and vegetables diet might be caused largely by potassium, judgment from the results of potassium supplementation in hypertensive or in persons with low customary K intake. Other foods and nutrients: other than testing a diet high in fruits and vegetables, the DASH study wasn't designed to work out different specific foods that scale back vital sign (Sacks et al., 1999). Compared with the fruits and vegetables diet, the DASH diet had a lot of vegetables, low-fat dairy farm merchandise, and fish, and was lower in white meat, sugar, and refined carbohydrates. Consequently, it had been higher in macromolecule, complicated super molecule, and atomic number 20, and was lower in sugar, saturated and mononnsaturated fatty acids, total fat, and cholesterol (Sacks et al., 1999).

Effectiveness of DASH diet

A study conducted in 1999 with the population of the study 459 healthy men and women with min age 44 years old. Average blood pressure was 132/85 mmHg. Among them 29% had mild hypertension. There are 60% populations of the sample covered by African American. 95% participant attended the DASH diet meal and this lowered blood pressure significantly, within a quick time around 2 weeks (Sacks et al., 1999). Compared with the control diet, the DASH combination diet significantly reduced blood pressure by 5.5 mm systolic and 3.0 mm diastolic using measurements made in the clinics, and 4.5 mm systolic and 2.7 mm diastolic for 24-h ambulatory measurements (Sacks et al., 1999). Among African Americans with hypertension, the DASH combination diet reduced blood pressure by 13.2/6.1 mmHg (Sacks et al., 1999).

From another case control study with a participant of 512 participants, 208 for DASH diet group and 204 for the control group. Around 95% of DASH diet group and 94 % of control group participated. Blood pressure measurement was done during each intervention period. Mean urinary sodium level was 142mmol per day in average during the high sodium period and 107mmol per day during the intermediate sodium period, and 65mmol per day during the low-sodium period. The levels of urinary potassium, phosphorus, and urea nitrogen (reflective of the intake of fruit and vegetables, dairy products, and protein, respectively) were higher in the DASH-diet group than in the control-diet group, and were nearly identical for all three sodium level (Sacks et al., 2001). The DASH diet, as compared with the control diet, resulted in a significantly lower systolic blood pressure at every sodium level and in a significantly lower diastolic blood pressure at the high and intermediate sodium levels. It had a larger effect on both systolic and diastolic blood pressure at high sodium levels than it did at low ones ($P<0.001$ for the interaction) (Sacks et al., 2001). As compared with the high-sodium control diet, the low-sodium DASH diet produced greater reductions in systolic and diastolic blood pressure than either the DASH diet alone or a reduction in sodium alone (Sacks et al., 2001).

Discussion

This review produced several key findings that are important for the prevention and treatment of hypertension. First, prevalence on hypertension is higher than in African American rather than any other group. They are in likelihood of developing hypertension at the young age.

Second there are many treatment and method for preventing and controlling hypertension. Lifestyle modification, change in the diet helps very much in controlling hypertension.

Third DASH diet lowered blood pressure at high, intermediate, and low levels of sodium intake, confirming and extending the findings of the previous DASH study.4 Thus, the benefits of following the DASH diet have now been shown to apply throughout the range of sodium intakes, including those recommended for the prevention and treatment of hypertension. Second, blood pressure can be lowered in the consumers of either a diet that is typical in the United States or the DASH diet by reducing the sodium intake from approximately 140 mmol per day (an average level in the United States) to an intermediate level of approximately 100 mmol per day (the currently recommended upper limit1), or from this level to a still lower level of 65 mmol per day. Moreover, reducing the sodium intake by approximately 40 mmol per day caused a greater decrease in blood pressure when the starting sodium intake was already at the recommended level than when it was at a higher level similar to the average in the United States. These results provide a scientific basis for a lower goal for dietary sodium than the level currently recommended.

References

1. APPEL, L. J., BRANDS, M. W., DANIELS, S. R., KARANJA, N., ELMER, P. J. & SACKS, F. M. 2006. Dietary approaches to prevent and treat hypertension a scientific statement from the American Heart Association. *Hypertension,* 47, 296-308.

2. BREWSTER, L. M., MAIRUHU, G., BINDRABAN, N. R., KOOPMANS, R. P., CLARK, J. F. & VAN MONTFRANS, G. A. 2006. Creatine kinase activity is associated with blood pressure. *Circulation,* 114, 2034-2039.

3. CAMPBELL, N. & CHOCKALINGAM, A. 1995. Prevention and control of high blood pressure: challenges and opportunities. *CMAJ: Canadian Medical Association Journal,* 152, 1969.

4. COOPER, R. & ROTIMI, C. 1997. Hypertension in blacks. *American Journal of Hypertension,* 10, 804-812.

5. FORD, E. S., GILES, W. H. & DIETZ, W. H. 2002. Prevalence of the metabolic syndrome among US adults. *JAMA: the journal of the American Medical Association,* 287, 356-359.

6. HALL, J. E. 2010. *Guyton and Hall Textbook of Medical Physiology: Enhanced E-book,* Elsevier Health Sciences.

7. NATIONAL HEART, L., AND BLOOD INSTITUTE (NHLBI). 2012. *What Is the DASH Eating Plan?* [Online]. Available: http://www.nhlbi.nih.gov/health/health-topics/topics/dash/ [Accessed 26 November 2013].

8. SACKS, F. M., MOORE, T. J., APPEL, L. J., OBARZANEK, E., CUTLER, J. A., VOLLMER, W. M., VOGT, T. M., KARANJA, N., SVETKEY, L. P. & LIN, P. H.

1999. A dietary approach to prevent hypertension: a review of the Dietary Approaches to Stop Hypertension (DASH) Study. *Clinical cardiology,* 22, 6-10.

9. SACKS, F. M., SVETKEY, L. P., VOLLMER, W. M., APPEL, L. J., BRAY, G. A., HARSHA, D., OBARZANEK, E., CONLIN, P. R., MILLER, E. R. & SIMONS-MORTON, D. G. 2001. Effects on blood pressure of reduced dietary sodium and the Dietary Approaches to Stop Hypertension (DASH) diet. *New England Journal of Medicine,* 344, 3-10.

10. STEFFEN, P. R., MCNEILLY, M., ANDERSON, N. & SHERWOOD, A. 2003. Effects of perceived racism and anger inhibition on ambulatory blood pressure in African Americans. *Psychosomatic Medicine,* 65, 746-750.